*A team of eight oxen extracting timber, a detail from 'The Vale of Ashburnham', a painting by J. M. W. Turner, 1816.*

# WORKING OXEN

## Martin Watts

## Shire Publications Ltd

# CONTENTS

Cover: *'A Gloucestershire Team of Oxen', from an original drawing by R. Hills at the Gloucester Folk Museum.*

ACKNOWLEDGEMENTS
I would like to thank John Creasey, Jonathan Brown at the Rural History Centre, Reading, Diana Zeuner, Frank Dean, Glynn Hartley, Bob Powell, John Gibson, Charles Martell, Alison Sinclair, Iona Joseph, Craig Barclay, C. Higginson and J. A. Sharwood & Co Ltd, Alastair Massie at the National Army Museum, Peter Brears, David Beevers, Sue and John Johnston, Sheri Steele, Andrew Hill, Jen Wilson-Kent, Richard Bradley, John Gall, the staff at the Craven Museum, Skipton, the staff at Kirkbymoorside Library, Dorothy Ellison, Maureen Wass, Helen Mason and the volunteers at the Ryedale Folk Museum, and finally Kirsty, Will and Emma.
   Illustrations are acknowledged as follows: British Library, page 7 (bottom); British Museum, page 1; Frank Dean, pages 13 (top), 16 (top), 27 (bottom); Frederick Remington Art Museum, page 19 (bottom); Gloucester Folk Museum (Gloucester City Council Cultural Services), cover; C. Higginson and J. A. Sharwood & Co Ltd, page 15 (bottom); Iona Antiques, London, page 11; Charles Martell, page 24 (centre); Colin Mason, page 24 (top); Royal Pavilion Libraries and Museums, page 19 (top); Ryedale Folk Museum, pages 13 (bottom), 14, 16 (centre), 25 (top two); J. Sinfield, page 4 (bottom); Sussex Life, page 26 (bottom); Yale Center for British Art, Paul Mellon Collection, page 30 (bottom); York Castle Museum, page 9; Yorkshire Museum, page 6 (top).

British Library Cataloguing in Publication Data: Watts, Martin. Working oxen. – (Shire album; 342). 1. Oxen – Great Britain – History. 2. Draft animals – Great Britain – History. 3. Agriculture – Great Britain – History. 4. Great Britain – Rural conditions. I. Title. 636'.0882. ISBN-10 0 7478 0415 X. ISBN-13 978 0 74780 415 4.

*Published by Shire Publications Ltd, Midland House, West Way, Botley, Oxford OX2 0PH, UK.*
*(Website: www.shirebooks.co.uk)*
*Copyright © 1999 by Martin Watts. First published 1999. Transferred to digital print on demand 2011.*
*Shire Library 342. ISBN 978 0 74780 415 4.*
*Martin Watts is hereby identified as the author of this work in accordance with Section 77 of the Copyright, Designs and Patents Act 1988.*

Printed and bound in Great Britain.

# THE HISTORY OF THE WORKING OX

An ox is a castrated male bovine at least four years old and trained to work. Oxen were used all over Britain for thousands of years and have helped to shape the land. They are an important but often neglected part of the nation's history. This book provides an introduction to working oxen and their history.

The story of working oxen begins with the aurochs (*Bos primigenius*), the ancestor of all domestic cattle. The cave paintings of France and Spain, between ten thousand and thirty thousand years old, are the first pictorial evidence of the aurochs. They show a large, alert, active beast standing over 5 feet (1.5 metres) at the shoulder, about 1 foot (30 cm) taller than most present-day domestic cattle. Both sexes had large forward-pointing upturned horns. Although the palette of the cave artists was limited, the red ochre and red brown that they used for the cows and the darker brown and black for the bulls seem both likely and appropriate. Evidence from later sources suggests that they had a pale eel stripe running from the forehead to the tail. The paintings show a whole range of activities: a bull nosing the rump of a cow; beasts with their heads and tails up in alarm, galloping, jumping or just strolling. They never present an image of docility or heaviness of movement, and Julius Caesar confirms this impression. He describes the aurochs in his campaign memoirs:

> The third species are those which they call 'uri'. These are little smaller than elephants with the appearance, colour and shape of a bull. They are strong and very fast and do not spare either man or beast once they have seen them. They [the Gauls] catch them in pits, which they have made with great care, and kill them. Youths harden themselves by doing this work and practise this method of hunting. Those who have killed many of them gain great honour when they parade the horns in public. But they cannot be domesticated and tamed, even if taken when very small. (Translation by Kay Beuchars)

*Cave painting of an aurochs, c.10,000 BC, from Lascaux, Dordogne, France, illustrated on a French one-franc stamp.*

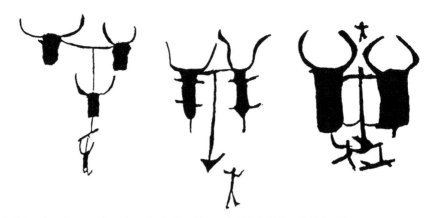

*Prehistoric rock engravings from the Italian Alps, c.1500 BC. (After C. Bicknell)*

Caesar had political reasons to enhance his account of the strength of his opponents, but, despite his tendency to exaggerate, the aurochs was a formidable beast with a reputation similar to that of the African wild buffalo of today. The cattle that Caesar would have been familiar with in Italy were the product of over two thousand years of domestication, but aurochs had continued to exist in the wild unchanged.

Despite Caesar's remarks about their ferocity, aurochs were tamed, and tamed aurochs became the first domestic oxen. The earliest evidence for domestication is found in the Middle East around ten thousand years ago. Exactly how they were domesticated is not known, but the animals must have been herded and managed as a food source. Then individuals would have been captured and tamed by a combination of skill, kindness and cruelty, starvation and dependence. The ancient civilisations of the Middle East, including the Egyptians and the Hittites, had successfully domesticated cattle while the people of Great Britain and the rest of

*Ploughing with oxen at Kom Ombo, Egypt, in 1997.*

4

*A Babylonian seed-drill, latter half of the second millennium BC. The peoples of the Middle East were among the first to domesticate the ox. (After 'Antiquity', 1936)*

Europe were still nomadic hunter-gatherers. The excavation of a neolithic summer hunting camp at Starr Carr in Yorkshire, dating from *c*.7500 BC, produced evidence that the aurochs was an important prey item for these early people.

As well as wild aurochs, archaeological evidence points to the existence of a separate domestic ox in the neolithic period. It closely resembled the aurochs but was smaller. It is thought that early neolithic settlers in Britain brought domesticated dogs, sheep and goats, but it is not clear whether they brought cattle too or domesticated the indigenous population of aurochs. They may have done both: the archaeological record is complex, with some bone remains closely resembling the aurochs and others the neolithic ox. The likelihood of interbreeding between wild and domestic stock and the strong possibility that domestic stock escaped and reverted to the wild also confuse the picture.

*The Golden Lyre of Ur (now in Iraq), c.2600 BC, in the Iraq Museum, Baghdad. The Sumerians worshipped a moon god, the young bull of heaven.*

An ox team depicted on a Roman coin of AD c.80.

A small Roman bronze found at Piercebridge, County Durham, gives a good idea of the costume of a Romano-British plough-man. (After Wright, 1875)

The Celtic people living in Great Britain in the bronze and iron ages imported a smaller breed of cattle, *Bos longifrons,* so named because it has a proportionally longer front part of the skull. The process of domestication continued, and these imported cattle would have interbred with the larger neolithic oxen. By the time of the Roman occupation the ox was firmly established as the most important traction animal on the farm. Improvements in farming techniques and the use of the wheeled plough pulled by oxen allowed more and more land to be farmed.

After the Norman conquest of 1066, William the Conqueror sent out commissioners to travel around his new kingdom, and they recorded its wealth in the Domesday Book, completed in 1086. They found Saxon agriculture almost totally geared to the ox. Only a minority of working farm animals were horses or mules; the majority were oxen, and the commissioners estimated the size and value of land holdings in terms of the ox. The number of oxen on a holding was a good indication of its potential productivity and consequently of the amount of tax that could be levied on it. Land was reckoned in terms of the amount a notional team of eight oxen could work in a year, the *carucate* (from the Latin *carruca,* a plough). The actual area of a carucate ranged from 10 to 18 acres (4.0 to 7.3 hectares), varying from region to region, depending on the soil and geographical conditions, but it was a very accurate method of determining productivity. Smaller holdings were measured in eighth parts of a carucate, called *oxgangs* or *bovates,* equivalent to what a single ox could work in a year. The commissioners recorded the livelihoods of men and women and their dependence on the ox, for example:

In Burdale, Ingifrithr, (she has) 10 bovates, taxable land for 4 oxen.
In Copgrove, Thorsteinn, Erneis' man, has there 1 plough.

Medieval agriculture with its communal open fields and individual strips of land

*A Roman ox cart with solid disc wheels.*

*A four-ox team ploughing in withers yokes, from the Luttrell Psalter in the British Library, c. 1340.*

*Fallen snow shows up the old ridge and furrow with its characteristic slow bend.*

evolved with the ox as the primary power source. The ownership of oxen increased the ability of people to contribute to the manorial system of agriculture with its rights, dues and co-operative practices. The farmers needed each other's help to work their strips of land successfully, as well as having to work for their feudal lord:

> William Bondi holds 14 acres [5.7 hectares].... With his three neighbours with their oxen joined together shall plough [for his lord] for one day at the winter sowing worth 1d and he ought to harrow for one day at the winter sowing worth 1d. He shall plough at Lent. (From the inquest on Baldwin Wake, 1282)

In Wales rules to govern this co-operation were set out in the Venedotian code. It carefully apportions benefits and responsibilities to all involved:

> The driver is to yoke in the oxen carefully, so that they be not too tight, nor too loose; and drive them so as not to break their hearts: and if damage happen to them on that occasion, he is to make it good; or else swear that he used them not worse than his own.

The size of the strips, their length and configuration had all evolved with oxen. In many parts of England the elongated S–shaped ridges and furrows that are the result of the oxen's work can still be seen. The field strips needed to have the shape of an elongated S because the long plough teams had to begin to make their turn well before the end of the furrow. A team of four oxen and a plough could be up to 40 feet (12 metres) long. The use of oxen governed a whole way of thinking: even the system of measurement was derived from them. A *furlong* was the length of a furrow that a team could reasonably be expected to pull without a pause. One furlong (220 yards, 201 metres) comprises 10 chains; one chain (22 yards, 20.1 metres) equals 4 poles. The *pole* (also called a *rod* or *perch*) was the length of the driving pole that drivers used to urge on the animals. Different parts of the country had different soils, traditions and topography and had different measurements too. In 1813 T. Davis, in his *Agriculture of Wiltshire*, recorded three alternative lengths for a pole: 15 feet (4.57 metres), 16 feet 6 inches (5.03 metres) and 18 feet (5.49 metres).

Despite the enclosure of land and the growing awareness that farming could be

8

*An eighteenth-century plaster frieze in the York Castle Museum. The oxen are urged on by a stubble boy. A small dog (almost worn away) rushes about above the plough.*

the subject of study and improvement rather than just a repetition of old and trusted ways, the ox continued to play a major part in farming and transport into the eighteenth century. Young lads like seventeen-year-old Henry Pinder were paid 34 shillings and 4 pence a year as stubble boys to help with the ox ploughs on Henry Best's farm in Driffield, East Yorkshire, in 1642. The Bests used both horses and oxen for draught although the heaviest work was reserved for the oxen. Henry Best complained that the haulage of huge loads of timber from Hull to the farm 'had almost broken all our waines [heavy ox carts] in comminge of 5 miles'. Roads were passable for very heavy loads only in summer, and Daniel Defoe records the problems in Sussex in 1742:

> I have seen one tree on a carriage, which they call there a tug, drawn by two and twenty Oxen; and even then 'tis carry'd so little a way, and then thrown down, and left for other Tug to take up and carry on, that 'tis two or three years before it gets to Chatham; for once the rains come in it stirs no more that year.

He goes on to record that near Lewes, in Sussex,

> I saw an antient Lady, and a lady of very good Quality I assure you, drawn to church in her coach with six oxen, nor was it done but out of mere Necessity, the Way being so stiff and deep, that no Horses could go in it.

Throughout this period the oxen were gradually being displaced by horses. Many draught teams used a combination of pairs of oxen led by a horse or horses. William Marshall, writing in 1790, records that in the Vale of Pickering, in North Yorkshire, 'from Time immemorial 4 to 6 oxen in yokes led by 2 horses also doubled was the invariable draught team of the country not only upon the road but in breaking up a fallow.' (Breaking the fallow was notorious as being the hardest ploughing work on the farm.) But, by 1790, 'throughout the Vale there is not perhaps a single ox employed in tillage', although they were still used on the road.

As Britain prospered, the population expanded, and farming was slowly transformed to meet the increased demand for food. Landowners were important and influential people, and it was both fashionable and profitable for them to be interested in farming and estate management. All the old practices and methods of husbandry came under close scrutiny. The general decline in the use of oxen continued, although it was revived at this time by some of the new fashionable and profit-conscious landowners and farmers. They followed the advice of some of the great farming commentators and either retained their oxen or reintroduced them to their farms and estates. A typical inventory of a gentleman farmer, Ralph Ward, in Cleveland in 1759 lists twelve horses worth £40 and fourteen oxen worth £140. His total estate was valued at over £21,000. The shift away from oxen to the use of horses was not universal, and old practices persisted. His contemporary John Huntley,

9

*A pair of French oxen in head yokes pull a heavy roller. (From 'Animalia' by Louis Figuier, fourth edition 1890)*

*A six-ox team ploughing on the South Downs in 1897. Scenes like this would have been familiar to farmers for centuries. (From 'Country Life Illustrated', May 1897)*

*'Three Oxen Yoked to a Plough', a painting by Daniel Clowes, c.1800. These are possibly Devon or Sussex oxen and have full leather harness as recommended by William Marshall (1790). He championed the use of oxen and of smart harness that would not only be more efficient but also enhance their status.*

living on a hill farm in the same district and working barely 20 acres (8.1 hectares), used oxen worth just £6.

The debate about the respective merits of horses and oxen for farm work had raged for centuries. As early as the thirteenth century Walter of Henley wrote: 'if the ground be tough your oxen shall werke where your hors shall stande styll.' In 1534 Master Fitzherbert, while allowing the greater speed of horses in light ground, still came out in favour of the ox, declaring roundly that even when the ox 'waxe olde, broysed, or blinde, for two shillings he may be fedde, and then he is mannes meate, and as good or better than ever he was, and the horse, whan he dyethe, is but caryen, and therefore me semeth, all thynges consydered, the ploughe of the oxen is moche more profytable than the plough of the horse'. The arguments in favour of the ox hinged on the cheaper feeding costs and its greater value when it came to the end of its working life. As William Marshall wrote in 1778, 'As an agricultural beast of labour he [the horse] has strength, alertness and tractability; but he is expensively purchased, expensively fed and rendered totally worthless by age or infirmity.' Robert Bakewell, the founder of modern livestock breeding methods and a highly influential and successful pioneer of new agricultural practices, still used his young Longhorn cattle in harness for carting jobs on his estate. King George III, a very keen farmer, changed his farm horses at Windsor for Glamorgan, Hereford and Devon oxen. A hard-headed northern businessman, George Culley, used oxen extensively on his Northumbrian estates.

Nevertheless the decline continued, and Robert Brown, in his report to the Board of Agriculture in 1799, wrote: 'Except in remote uncultivated parts, there is hardly an ox team employed, unless it be on the farms of landed proprietors, who probably have been induced to use them from public spirited motives, without enquiring into

'The Lancashire Ox', by Thomas Bewick, from the 'History of Quadrupeds', 1790. In the background can be seen two oxen and a horse ploughing together.

the practical result of their operations.'

The size and strength of horses had improved, and the difference between the speeds of teams of oxen and teams of horses had increased. There was a growing number of improved roads with hard surfaces, which suited the hooves of horses rather than those of oxen. Much farm work could be done using fewer horses. In the nineteenth century new machines like the reaper could not be pulled fast enough by oxen. Traditionally oxen were worked for only a part of the day. All cattle need a substantial bulk of food for their digestive systems to operate properly, and it takes some time for them to eat this; by contrast, a horse functions well on a more concentrated high-energy food that can be fed at the start and finish of the day's work. Paradoxically, this relatively short period of daily work should have been better suited to smaller family-run farms with less acreage to work, but the versatility of the horse often proved a decisive factor. Mrs Bennet in Jane Austen's *Pride and Prejudice* was often anxious to borrow horses for her carriage and thus interrupt the work on the farm. In the end it was often the larger estates, which could afford to maintain a variety of animals for different purposes, that kept a team of oxen.

Half-bred Shorthorn oxen preparing to harrow in Oxfordshire in 1897. (From 'Country Life Illustrated')

*Mr Pelling of Saddlescombe with his team of four Sussex oxen. (By courtesy of Mr Frank Dean, who was given the original photograph by Mr Pelling's grandson.)*

*Two oxen with a heavy wagon, twenty-two people with rakes and forks and one horse pulling a sweep, all getting in the hay on the Earl of Feversham's estate, Duncombe Park, North Yorkshire.*

The great breeders of livestock at this time mainly concentrated on developing cattle as producers of milk and beef and paid relatively little attention to the very different physical qualities needed in a good draught ox. A draught ox needs strong forequarters, plenty of muscle and no fat; beef animals need large hindquarters, and in those days 6 inches (15 cm) depth of fat along the back was considered highly desirable. The contemporary portraits of prize livestock show animals mostly quite unsuited to physical work.

There were exceptions to this overall trend in cattle breeds, chief among them being the Hereford, the Devon Red and the Sussex. The Devon Red and the Sussex both came from areas where the tradition of working oxen was still strong; Sussex was their last real stronghold. The Hereford was developed as a dual-purpose animal by the Herefordshire farmers, who wanted a strong docile beast that they could put to work for a few years before it went to be fattened and slaughtered.

Above all the economic arguments, perhaps it was the sheer appeal of horses and the fondness of the British people for them that spelt the end for oxen. By the end of the nineteenth century working oxen were a rarity. They could still be seen in parts of Sussex, in Ayrshire and in north-east Scotland, while some enthusiasts like the Earl of Bathurst and the Earl of Feversham kept a team on as a reminder of past ways. They might still sometimes help out hard-pressed farmers and smallholders, who could always harness a docile cow or bullock, sometimes paired side by side with a single horse. Occasionally at harvest, when every type of vehicle and all help were needed, a quiet pair of bullocks would join in.

*The decline in the use of oxen is obvious in this celebratory photograph of the harvest home on a farm in northern England c.1900. A pair of oxen can be seen at the very end of the row of horses. The proud farmer holds a terrier, and his wife restrains a goat. Note also the magnificent haystacks.*

*In 1893 Hugon and Company began using colourful ox-drawn wagons to advertise their Atora suet all over Britain. It was one of the most successful and long-running publicity campaigns of its day.*

A few agricultural teams lingered on into the twentieth century; Lord Bathurst's in Gloucestershire continued until 1964. Up to *c.*1940 Atora suet was delivered in special wagons drawn by pairs of Hereford oxen.They were sold on to Chipperfield's to join the circus parade. The innovative academic reconstruction of iron age life by Peter Reynolds at Butser, Hampshire, in 1972 experimented with Dexter cattle in draught. Since then there has been a growing interest in working oxen. The break in

*The last 'Atora' oxen, two Herefords called Dick and Sailor, outside Hereford Cathedral.*

*Dick and Sailor outside the Abergavenny Arms at Rodmell, East Sussex, in 1933. Mr C. Dean, the farrier, looks on.*

*A few teams of oxen lingered on into the twentieth century. This carnival took place between the world wars.*

*Lancelot and Arthur, two fine Longhorns, being trained by John Johnston at Cruckley Animal Farm, near Driffield, East Yorkshire. They have featured in a number of films and television dramas.*

the tradition of working them has made it very difficult for people interested in learning how to work them to find the right equipment and techniques. However, despite these difficulties several teams have been trained. Charles Martell of Gloucestershire has spent many years working with oxen and has pioneered the rediscovery of the techniques and skills. He now has a team of six Gloucesters, the largest team in western Europe. Several rare breeds centres and farm parks have working pairs, including the Cotswold Farm Park in Gloucestershire and Cruckley Animal Farm near Driffield in East Yorkshire. It is now possible to see these magnificent animals in action once more.

*Charles Martell, a leading expert on draught oxen in Britain, with his magnificent team of six Gloucesters: from front to back, Will and Weaver, Buck and Benbow, Sport and Spider. They are totally controlled by the long pole and word of command.*

# OXEN AT WORK

The docility of the ox is legendary: this is due partly to castration and partly to the tendency of bovines to face a threat rather than run from it. Even when thoroughly alarmed, they will not normally bolt for considerable distances like a horse would. However, managing even a pair of oxen can call for strength and stamina. A team of six or eight, weighing up to a ton each, is hard work, and they must be treated with respect. During the wars in South Africa the soldiers in the transport corps did not understand that oxen were not the same as horses, and as a result they experienced considerable difficulties with them. Although some individuals may be difficult and flighty, they have a deserved reputation for gentleness.

Oxen are animals of habit and right from the start are trained to be on the left (nearside) or right (offside) of the team. The nearside animals, closest to the driver as he walked on the left as the team moved forward, had single-syllable names like Ben or Buck, and their offside partners had a two-syllable but similar name such as Benbow. When ploughing, the nearside animal (Ben) always worked on the land (unploughed) side and his partner (Benbow) on the furrow side. Individual animals get very attached to their working partner and their overall position in the team, and it is not easy to make any changes!

Ox teams of more than two animals usually needed at least two people to operate them, particularly if the job in hand was more complicated than simple road haulage. Typically, one person would operate or manage the implement, and another would walk alongside the animals to keep them going. Ælfric (c.995–1020) described the ploughman in early medieval England:

> I work hard; I go out at daybreak, driving the oxen to the field, and I yoke them to the plough. Be it never so stark winter, I dare not linger at home for fear of my lord; but having yoked my oxen and fastened share and coulter, every day I must plough a full acre or more. I have a boy driving the oxen with a goad iron, who is hoarse with cold and shouting. And I do more also. I have to fill the oxen's bins with hay and water them, and take out their litter.... Mighty hard work it is....

Walking in front and leading the oxen tends to slow them down or even to make them stop altogether. Highly skilled drivers, when working single-handed with a team, had to walk to one side to get the best out of them. Long whips or long straight

*A massive load of wool being hauled by fourteen oxen driven by just one man in Australia, c.1930. Men and women who earned their living driving teams of oxen were known as 'bullockies' in Australia.*

poles as well as goads were used to guide and encourage the animals. There were regulations that forbade the use of a nail longer than a barleycorn (8.5 mm) in the end of the pole. As with all domestic animals, a good working team is based on affection and trust as much as the whip. Australian drivers used long whips to steer the team. A good driver very rarely actually touched them with the whip; rather the animals, particularly the leading pair, were trained to respond to the position of the whip crack and the word of command. Welsh ox drivers in medieval times sang songs to encourage their teams. Although oxen were sometimes fitted with nose rings, these do not seem to have been used very often as a controlling mechanism when they were working.

In a large team the most important animals are the pair nearest the vehicle, known as the polers, and the leading pair. The polers had to be steady and strong, being the only members of the team that can help with braking the vehicle when going downhill. They have to take a great deal of the strain on their necks. The leading pair also have a crucial role: as well as setting the direction, they must be prepared to set a good pace and not be daunted when faced with a hard pull.

The rapport between driver and team could be considerable. There are instances of Australian drivers pulling out tree-stumps single-handed with teams of eight

*A grape festival in Italy in 1937. Only two oxen pull this huge float with sixteen passengers.*

18

*In Preston Manor, Brighton, is this painting showing the amazing and presumably unrehearsed co-operation and steadiness of eighty-six oxen working together. They pulled a windmill from Regency Square in Brighton 2 miles (3.2 km) to the Dyke Road on 28th March 1797.*

oxen. An obvious clue to the nature of this relationship is the lack of the many restraining and safety measures that are associated with draught horses. There were no blinkers, no bridle, no bit, no breechings, no crupper, no saddle, nor even reins, just a simple yoke and bow. The fact that none of these was necessary is a testament to the nature of a well-trained ox working in a withers yoke. The ox was not broken – it was induced to co-operate.

So much haulage work over the last two thousand years has been done by oxen. Until the advent of the canal and railway systems they were the main movers of heavy loads. They even played a part in the railway age. Thomas Fenton, a wealthy mine-owner, used them to pull the coal wagons on his railway at Lofthouse in West Yorkshire.

An ox can work for seven or eight hours a day before needing a rest. His natural working gait is a walk, and he can travel at a speed of 2 to $2^1/2$ miles an hour (3.2–4.0 km/h).

*'The Santa Fe Trade' by Frederick Remington. The covered wagons are pulled by oxen, and the cowboys are riding mules.*

19

# EXPORT CART FOR OXEN.

Fitted with Pole for Oxen, Portable Top Boards, Loose Tail Board, etc.

| No. | Carrying capacity. | Wheels | Tires. | Prices | Approximate measurement. | Approximate gross weight. |
|---|---|---|---|---|---|---|
| 8 | 1680-lbs. | 4-ft. 0-in. | 4 in. | as per list. | 75 cubic ft. | 650-lbs. |
| 9 | 2240-lbs. | 4-ft. 6-in. | 4-in. | ,, | 77 cubic ft. | 675-lbs. |
| 10 | 2800-lbs. | 4-ft. 6-in. | 4-in. | ,, | 80 cubic ft. | 700-lbs. |

GENERAL SPECIFICATION OF MATERIALS see page 3.

*An ox cart made by the East Yorkshire & Crosskills' Cart & Waggon Co Ltd for export abroad. Apart from the pole, there is little difference between this ox cart and horse carts. A number of fine ox carts still exist, notably at the Museum of Welsh Life at St Fagans, near Cardiff.*

Oxen are capable of pulling enormous weights and have achieved some notable feats. Near Brighton in Sussex in 1797 eighty-six oxen moved a windmill 2 miles (3.2 km) across country from Regency Square to Dyke Road. There are similar accounts from Australia of teams moving houses many miles. The American West was opened up with wagons drawn by mules and oxen more often than with the more photogenic horses of Hollywood films. In his flamboyant autobiography William Cody (Buffalo Bill) describes a typical wagon train in 1857:

> The wagons used in those days by Russell Majors & Waddell were known as the J. Murphy wagons, made at St Louis specially for the plains business. They were very large and were strongly built, being capable of carrying seven thousand pounds of freight each ... and drawn by several yokes of oxen.

### Oxen at war

For centuries oxen have hauled military supplies. Genghis Khan (c.1162–1227), the Mongol conqueror, used them to transport his massive tents. The British army used them in the Empire for hauling supplies and artillery. In India in 1896 they had three thousand oxen for draught and three thousand for artillery siege trains. The Bulgarians used buffalo as military draught animals in the First World War. As recently as 1933 oxen were still in use for artillery trains in India and Abyssinia (Ethiopia), and the Veterinary Department of the War Office produced standing orders and instructions on oxen for transport officers in the field.

Perhaps the most notorious use of oxen was in the Anglo-Zulu War of 1879 and the Boer War of 1899–1902. At the time the British army normally used mules or horses for wagon transport, but because of the long sea voyage it was thought expedient to hire local transport. However, there were not enough mules in South Africa and the Commander-in-Chief, Lord Chelmsford, had to purchase thousands of oxen and wagons to transport his army less than a hundred miles (160 km) to the

*Heavy artillery: a 40 pounder on parade in India, photographed by Frederick Bremner in the 1890s. These guns had both elephant and ox teams. The ox teams were considered steadier under fire.*

battle zone. Chelmsford began his invasion of Zululand in 1879, with just under seventeen thousand troops and a few thousand oxen. The soldiers had no experience of handling oxen, with disastrous consequences. Poor livestock management, lack of grazing, the slow speed and non-existent roads nearly brought the campaign to a standstill, as well as costing hundreds of thousands of pounds. Changing

*The invasion of Zululand in January 1879: Colonel Glynn's column crossing the Buffalo river near Rorke's Drift. (From 'Illustrated London News')*

*Troubles with oxen in South Africa, c.1900. Sketches by E. Caldwell, who went out to South Africa specially to sketch these pictures from life for the children's adventure story 'Jock of the Bushveldt' by Sir Percy Fitzpatrick.*

*Sir Samuel Baker searching for the source of the Nile. Both he and his wife rode oxen on their perilous journey. (From 'Albert Nyanza, Great Basin of the Nile' by S. Baker, 1885)*

tactics, he hired teams with local drivers and set up chains of supply depots, one of which was the famous Rorke's Drift. Before the end of the war over 27,000 oxen were working to keep the army supplied.

Twenty years later the British army was back in South Africa to fight the Boers. Again thousands of animals and wagons were used, particularly in the early stages before the railway was extended. The Commander-in-Chief, Field Marshal Lord Roberts, used the same system of hiring local teams, and before the end of the campaign there were over forty thousand oxen working for the British army, under the command of a major-general. The long marches and poor management of the animals again led to high mortality rates – over 11 per cent a month. The Boers, by contrast, were renowned experts in the use of oxen and in the early months of the war used them for moving supplies and artillery. Despite all the difficulties experienced by the British army in the use of oxen, when they were properly used the army did appreciate their potential. In his report to the Royal Commission on the War in South Africa Lord Roberts wrote:

> The load for a wagon was 6000 pounds [2722 kg]. The oxen were, as a rule, fine animals, and very tractable. The curious thing about them was that they would pull together, however large the team might be. With heavy guns as many as 20 spans of oxen were employed, and when they were on the move the trek chain was always taught.

22

# OX YOKES

Throughout the recorded use of oxen they were most frequently used in pairs or spans, linked together by a stout piece of timber called a *yoke*. The yoke not only paired the oxen but also increased the control over them, the weight and strength of one ox serving as a brake on the other. The training of a pair of oxen usually began at a very early age; they would learn and grow together. Working pairs could become very attached to each other and were often observed still grazing side by side after work.

The earliest recorded yokes were head yokes. Examples of head yokes for oxen from the bronze age are very similar to those still used in Europe today. These carefully shaped pieces of wood made use of the horns as strong fixing points. They were lashed to the head and horns with leather strips. The forehead was often protected by a pad of cloth or bundles of dry grass. Head yokes were very effective for controlling the animal but had a number of faults. They did not allow the ox to use his strength with maximum efficiency. When an ox pulls, he exerts his greatest strength with his head down, and in a sense he pushes rather than pulls. The head yoke forced the head up. It also transmitted all the jarring of the wheels when working on hard uneven surfaces straight to the head, causing unnecessary pain and discomfort. It also prevented the ox from shaking his head when irritated by flies. All these difficulties are solved by the shoulder or withers yoke. The ox can exert an even greater force using his withers or shoulders than he can with his head and neck. The head yoke was soon superseded in Britain by the withers or shoulder yoke, and by the fourteenth century the head yoke is hardly ever illustrated in manuscripts or carvings.

The withers yoke consists of a large piece of timber with two gentle bends shaped to fit the shoulders. Each animal then had a rope or more commonly a bent piece of wood, or *bow*, passing under his neck and attached to the yoke. The springy nature of the bow allowed it to be unfastened and slipped around the neck. The ox bow is all that stopped the ox from slipping out from under the yoke. There was sometimes

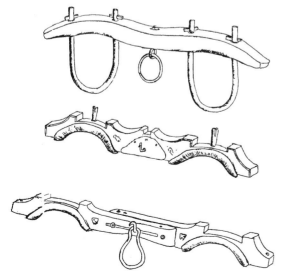

*(Top) A withers yoke complete with ox bows. (Centre) A continental head yoke. (Bottom) An adjustable head yoke.*

23

*Two oxen with their fine withers yoke at Old Sturbridge Village in Massachusetts.*

*A Sussex yoke and bows.*

*An old Australian yoke with iron bows at Alice Springs, Northern Territory. The bows have two positions for the keys to fit larger necks. The big centre ring is held to the yoke by a journal with another key. The keys could also be secured with green hide to stop them falling out.*

an upward tilt at the bottom of the bow to reduce any chance of it restricting the windpipe. At the centre of the yoke was either a large metal ring or hooks and a ring to which a pole for a cart or chains for an implement could be attached. The yoke and its bow are all the harness that is needed.

The simplicity and cheapness of yokes would have appealed to frugal farmers. The yoke and bows, unlike leather harness, did not need special craftsmen to make them. As Thomas Tusser suggested in 1557:

Yokes, forks and such other let bailiff spy out
and gather the same as he walketh about
and after, at leisure, let this be his hire,
to beath them and trim them at home by the fire.

Above left: *An oxbow stone at the Ryedale Folk Museum. It can be used to make three sizes of bow, evidence of the need to start training the animals at a young age.*

Above right: *A Yorkshire oxbow stone. The four peg holes can be clearly seen.*

Left: *An ox using a horse-type harness, photographed in the late nineteenth century. The collar has two straps that can be undone to open it up in order to get it past the horns.*

To 'beath' the wood is to heat it prior to bending it into position. In North Yorkshire there are a number of specially carved stones that acted as patterns and clamps for making ox bows. The hot timber, often ashwood in Britain, was bent around the raised-up U-shaped section of the stone and tied and pegged in place. Wooden bows were sometimes replaced by iron ones. Examples of yokes with iron bows have been found in Devon and Scotland, and iron was commonly used in Australia in the nineteenth century.

Although the withers yoke was more efficient and more comfortable, it caused some problems: in particular it restricted the movement of the shoulders and in wet weather could chafe. During the revival in the use of oxen, particularly among gentlemen farmers from the mid eighteenth into the nineteenth century, some of these farmers used leather harness. This was very similar to horse harness, complete with bridle, collar, belly band, even occasionally saddle, and breechings. However, ox harness has a different-shaped collar and hames with lower pulling points. If the ox had horns, the collars had to be hinged to allow them to be opened up and put round the animal's neck rather than slipping it over the nose. Modern oxen today are driven in both harness and withers yokes. The use of breechings with a leather harness allows the ox to slow down and brake the wagon more effectively when going downhill.

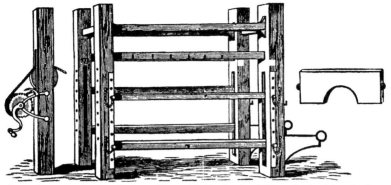

*A crush or trevis for shoeing. (From 'A Textbook of Horseshoeing' by A. Lungwitz, 1884)*

# OX SHOES

Not all working oxen were shod, but it was frequently done if they had to work on hard ground or were being driven long distances on the road. Drovers walking large numbers of cattle across the country often had them shod to keep their feet in sound condition and thereby ensure a good price at the market. Smiths living near the drove roads often had dozens of beasts to shoe at a time.

It seems odd that such a tractable animal as an ox should apparently require substantial restraint when it was being shod. A horse, when trained, will stand on three legs and allow the farrier to work on the fourth, flexing the lower leg both forwards and backwards. Oxen, it seems, were often totally immobilised.

Several methods were used. The first involved a *crush* or *trevis*. Four stout posts were set in the ground, and the ox was led between them. Then further ropes and timber were put in place to stop him moving. Some continental crushes had a broad belly band with a windlass that could be used to support his weight. Each foot in turn was then lifted up and either just rested on or lashed to a short stout post 12 inches (30 cm) or so from the ground. A number of such crushes can still be seen in continental villages, and there is a record of one in the Vale of Pewsey in Wiltshire.

The second method was to tie ropes to the ox's feet and gradually pull them

*Preparing to throw an ox before shoeing at Saddlescombe in Sussex. His front legs are held with a stout leather strap. The two men nearest the camera will gradually pull his front legs together, so forcing him to fall over. Note the two ox yokes propped on the hedge behind.*

*Once felled, the ox has all four feet tied together on to a tripod, and a boy sits on its neck.*

together until he fell over. This was called 'throwing the ox'. When he was down, all four feet were tied together. A picture of shoeing an ox in Sussex shows the hooves tied to a tripod at a convenient height for the smith to work on them. Sometimes the feet were tied in pairs, with the two fore feet together, the two hind feet together and a long pole between them.

A third method required two strong assistants. The animal was tied tightly by the head to a stout gate or post. When each back foot was being

worked on, it was supported on a pole. To work on the fore feet, they were held tight, one at a time, with a rope. Mr Frank Dean recalls shoeing a Hereford bull in the early 1960s with this method. The great scarcity of shoeing crushes or trevises in Britain might suggest that very many oxen were shod in this last way.

*It is possible to shoe an ox without a crush or throwing the animal; a stout pole is used, with two strong men to hold it up.*

An ox needs eight shoes for a full set, two to each cloven hoof. Ox shoes are sometimes known as cues, ques or kews. There were a great variety of shoes. Some are shaped like a comma, with the top of the comma at the front of the foot; others are a bent oval or even a triangular plate covering the whole claw. They are usually about a quarter of an inch (6 mm) thick or less. Those for the front feet are normally a little bigger than those for the rear.

They are nailed, using between three and

*Frank Dean of Rodmell, Sussex, shoeing a Hereford bull to straighten his bow legs. He used a pole to support each rear leg in turn.*

27

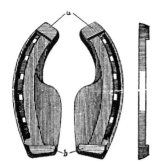

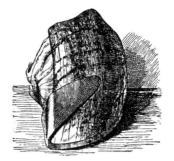

(Left) A shod ox claw in a slipper-shaped shoe, and (far left) a pair of machine-made shoes. (From a 'Textbook of Horseshoeing' by A. Lungwitz)

Ox shoes. (Top, left to right) A wedge-shaped shoe, from Beamish, the North of England Open Air Museum; a bent oval shoe with circular countersunk holes, from Craven Museum, Skipton; a comma-shaped shoe with a fullering groove, from the York Castle Museum; (bottom, left to right) a slipper-shaped shoe with additional tongue to fit over the top of the claw, from the Ryedale Folk Museum; a shoe with toe clip and turned-up rim at the heel, and a small shoe with a turned-up rim, both from the Craven Museum.

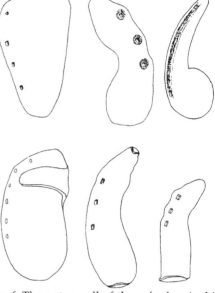

six nails, into the outer edge of the hoof. The outer wall of the ox's claw is thinner than a horse's hoof, and the nails were shorter than horseshoe nails – about an inch (25 mm) long. Smiths would sometimes grease the nails before hammering them in. Keeping the shoes on the feet was a problem, and several developments were made. Some shoes had toe clips, an additional tongue of metal that turned over the tip of the hoof at the front, as well as a raised ridge at the heel of the shoe. This reduced the likelihood of casting a shoe. Some smiths used a nail shaped like a hockey stick with a large flat head; it stretched across the shoe and provided additional grip on the road surface, as well as reducing the chances of the shoe being cast. The most radical solution was a shoe with a tongue of metal extending from its outside front edge right over the top of the claw, like a slipper. This type is common on the continent of Europe.

Shoes with a fullering groove were fitted with a more conventional nail with a smaller head. Shoes with calkins to enhance the grip on the road have been found.

Glynn Hartley from Alice Springs, who worked as a farrier in the Northern Territory of Australia for many years after the Second World War, recalls that, when fitting the shoe, 'the hoof is levelled at the toe. The heel very rarely needs touching and the que [shoe] was nailed on exactly like a horse ... Care must be taken that it's not possible for the que on one toe to rub against the other toe.'

*This magnificent ox house on a farm in Lancashire is a Grade II\* listed building and is still in use for housing cattle. The earliest written record of its existence dates from 1670, but it is probably much older.*

# OX HOUSING

Very few specialised agricultural buildings survive from before the eighteenth century, and very few of those housed oxen. Medieval peasants, if they owned an ox, would most likely have shared their simple one-roomed house with it in the winter. During the summer months it would have remained outside. An outstanding example of an ancient ox house in Lancashire shows how important the animals were in the economy of the farm. It is a very large building. The side that faces the prevailing weather is all brick with a few ventilation holes; at ground-floor level the other three sides are timber-framed with panels of wattle and lime plaster. The second storey is jettied out from the lower one and supported on massive tie beams. It is clad with timber boards. Inside on the ground floor there are six rows of low stalls or partitions running across the building, providing housing for thirty-six animals. The oxen were tied up in the individual stalls, each with its own manger. Mucking out was efficiently done along the rows, and the manure was thrown out through the doors at the end of each one. The doors themselves are over 3 feet 6 inches (107 cm) wide to allow for the large size of the ox and its often widespread horns. Hay for winter feed was stored in the huge hayloft running the entire length of the building. To get to the loft, the farmer climbed up a rough spiral staircase. There are pitching eyes in both ends and along one side. This building is most probably an exception rather than the norm.

In the seventeenth and eighteenth centuries farmers would still have stalled their oxen over winter, but in a smaller building that has often had the characteristic internal fittings altered to accommodate horses. The manger and dividing partitions for horses are higher than those for oxen. Horses were usually tied to a strong metal

ring whereas cattle, both cows and oxen, were often tied to a ring that slid up and down a vertical bar. Prior to the widespread use of horses the word 'stable' could refer to housing for an ox as well as that for a horse. However, in surveys of the Duncombe estate in North Yorkshire, in the seventeenth and eighteenth centuries, ox houses and stables are differentiated on paper, although it is now very difficult to find any traces that conclusively prove the use of a particular building. Another common form of housing used for centuries by small-scale farmers was a simple pole barn under a thatched roof. The roof was often thatched with an unthreshed crop, to keep the crop out of reach of vermin. It seems that in this type of housing it was the usual practice to bed the animals down on layers of bedding and create useful manure, rather than cleaning them out daily.

*'Cow Lodge with a Mossy Roof', painted by Samuel Palmer in Kent in 1828. Shelters like this for cattle were once common throughout Britain.*

*An ox team at Helmsley, North Yorkshire, c.1900.*

# CONCLUSION

Despite having disappeared from the rural scene, the ox has left a huge legacy behind. As well as the physical remains, the ridges and furrows and a system of linear measurement that seemingly defies logic and taxed the patience of schoolchildren for centuries, there are the names of wild flowers: the ox-eye daisy, with its long white petals like eyelashes, the oxlip and the bristly ox-tongue. One can still buy shoe polish of a rich red colour called 'ox blood'. Geographers call the characteristic U-bends in a river 'ox bows'. The words 'goad' and 'yoke', although no longer used in agriculture, are still in current use as metaphors. The names of many towns and villages remind us of the existence of oxen, including Oxted and Oxford, and even in London road names recall a possible earlier use: Oxestalls Road (SE8), Oxgate Lane (NW2). People are said to be as slow, gentle, patient or strong as an ox. And at Christmas it was the ox that made room for Mary and Joseph.

# FURTHER READING

Braden, L. *Bullockies*. Seal Books, 1968.
Conroy, Drew, and Barney, Dwight. *The Oxen Handbook*. Butler Publishing, 1986.
Creasey, John. 'The Draught Ox', *Heavy Horse*, March 1977.
Fenton, Alexander. *Scottish Country Life*. John Donald, Edinburgh, 1976.
Fenton, Alexander. *The Shape of the Past: Essays in Scottish Ethnology*. John Donald, Edinburgh, 1985.
James, Terry, and Anderson, Frances. *In Praise of Oxen*. Nimbus Publishing, 1992.
Jenkins, Geraint (editor). *Studies in Folk Life: Essays in Honour of Iorweth C. Peate*. Routledge & Kegan Paul, 1969.
Jenkins, J. Geraint. *Agricultural Transport in Wales*. National Museum of Wales, Cardiff, 1962.
Langdon, John. *Horses, Oxen and Technological Innovation*. Cambridge University Press, 1986.
Martell, Charles. 'Draught Oxen', *Heavy Horse World*, 1994.
Moncrieff, Elspeth, and Joseph, Stephen and Iona. *Farm Animal Portraits*. Antique Collectors' Club, 1996.
Piggott, Stuart. *The Earliest Wheeled Transport*. Thames & Hudson, 1983.
Piggott, Stuart. *Wagon, Chariot and Carriage*. Thames & Hudson, 1992.

# PLACES TO VISIT

*Aberdeenshire Farming Museum,* Aden Country Park, Mintlaw, Peterhead AB42 5FQ.
   Telephone: 01771 622906. Website: www.aberdeenshire.gov.uk/museums
*Angus Folk Museum,* Kirkwynd Cottages, Glamis, Angus DD8 1RT.
   Telephone: 01307 840288. Website: www.nts.org.uk/Property/5
*Beamish: The Living Museum of the North,* Beamish Hall, Beamish, County Durham DH9 0RG. Telephone: 0191 370 4000. Website: www.beamish.org.uk
*Bollin Valley Partnership,* County Offices, Chapel Lane, Wilmslow, Cheshire SK9 1PU.
   Telephone: 01625 534790. Website: www.cheshireeast.gov.uk
*Cotswold Farm Park,* Guiting Power, Cheltenham, Gloucestershire GL54 5UG.
   Telephone: 01451 850307. Website: www.cotswoldfarmpark.co.uk
*Craven Museum,* Town Hall, High Street, Skipton, North Yorkshire BD23 1AH.
   Telephone: 01756 706407. Website: www.cravendc.gov.uk
*Gloucester Folk Museum,* 99-103 Westgate Street, Gloucester GL1 2PG.
   Telephone: 01452 396868. Website: www.gloucester.gov.uk/folkmuseum
*Home Farm Temple Newsam,* Leeds, West Yorkshire LS15 0AD. Telephone: 0113 264 5535.
*Museum of English Rural Life,* 6 Redlands Road, Reading, West Berkshire RG1 5EX.
   Telephone: 0118 931 8660. Website: www.reading.ac.uk/merl
*Museum of Welsh Life,* St Fagans, Cardiff CF5 6XB.
   Telephone: 01222 573500. Website: www.museumwales.ac.uk
*Ryedale Folk Museum,* Hutton le Hole, North Yorkshire YO6 6UA.
   Telephone: 01751 417367. Website: www.ryedalefolkmuseum.co.uk
*Weald and Downland Open Air Museum,* Singleton, West Sussex PO18 0EU.
   Telephone: 01243 811363. Website: www.wealddown.co.uk
*York Castle Museum,* Eye of York, York YO1 9RY.
   Telephone: 01904 687687. Website: www.yorkcastlemuseum.org.uk

Printed and bound by CPI Group (UK) Ltd, Croydon, CR0 4YY

11/10/2024

01043558-0002